Canon EOS R5 Mark II Guide

I0062454

The Easy Guide to Pro Photography and Video with the R5 Mark II Mirrorless Digital Camera

Georgette Howard

Table of Contents

Copyright © 2025 by Georgette Howard

All rights reserved. No part of this publication may be reproduced, distributed, or transmitted in any form or by any means, including photocopying, recording, or other electronic or mechanical methods, without the prior written permission of the publisher, except in the case of brief quotations embodied in critical reviews and certain other non-commercial uses permitted by copyright law.

Introduction

Unleash the Power of the Canon EOS R5 Mark II, even If you've never held a camera before.

Whether you're a total beginner, a hobbyist ready to level up, or a working creative switching to Canon's most advanced mirrorless system, this guide is your fast track to mastering the Canon EOS R5 Mark II without the overwhelming jargon or guesswork.

The EOS R5 Mark II isn't just a camera. It's a hybrid powerhouse packed with cutting-edge technology 45MP stills, mind-blowing 8K video, AI-powered autofocus, and next-gen image stabilization. But let's be honest: with great power comes a steep learning curve. That's where this book comes in.

What This Book Does Differently?

Forget dense manuals and confusing YouTube rabbit

holes. This guide was built with YOU in mind, the real-world user who wants to shoot better, faster, and smarter. It simplifies everything, from unboxing to pro-level control, and walks you through every button, mode, and menu with clear explanations and real-life examples.

Here's What You'll Learn Inside:

Effortless Setup & Navigation: Get started fast with easy-to-follow steps for charging, memory card setup, and first-time menu customization.

Camera Anatomy Made Simple: Know what every button, dial, and screen does and how to use them like a pro.

Auto to Manual Mastery: Understand exposure, focus, and camera modes with "dummy-style" walkthroughs that actually make sense.

Genre-Based Shooting Guides: Tailored tips for

portraits, landscapes, wildlife, video, and more, learn to dial in the right settings every time.

Pro Workflow Tips: Learn how to organize, back up, and edit your files using Lightroom, Photoshop, and the camera's dual card slots.

Troubleshooting Fast: Fix common problems like a jammed lens, poor focus, or overheating without panic.

Bonus Visuals & Diagrams: Easy-to-read graphics that explain concepts like white balance, histograms, and autofocus areas at a glance.

Who This Book is For?

- ✓ New camera owners looking for a confidence boost
- ✓ DSLR switchers moving into the mirrorless world
- ✓ Content creators, vloggers, and YouTubers
- ✓ Seasoned pros needing a fast-reference upgrade guide

✓ Anyone who wants to get the most out of the R5 Mark II without wasting hours online

Your Creative Journey Starts Now!

The Canon EOS R5 Mark II is the most advanced tool Canon has ever built but it's your vision, skill, and workflow that bring it to life. This book makes sure you can capture stunning photos and cinematic video, no matter your starting point.

If you're ready to ditch the guesswork and start creating like a pro, this guide is your new favorite camera companion.

PART 1: GETTING STARTED – THE DUMMY-PROOF FOUNDATION

Chapter 1

Welcome to the R5 Mark II

Who This Guide Is For?

Whether you're holding the Canon EOS R5 Mark II in your hands for the first time or you're upgrading from an earlier model and wondering what's changed — this guide is for you.

This book is written for:

- Complete beginners who feel overwhelmed by the buttons, dials, and menus but are excited to learn.

- Hobbyists who want to break out of Auto mode and start shooting like pros.

- Aspiring professionals who want to unlock the full potential of this high-performance mirrorless camera.

- Content creators looking to step up their photo and video game with Canon's latest technology.

If any of that sounds like you, welcome! You don't need to be a tech genius or a camera wizard to master this device. You just need curiosity, a bit of patience, and a willingness to experiment. I'll take you through everything, one clear step at a time.

What Makes the R5 Mark II Special

The Canon EOS R5 Mark II is not just another camera, it's a **professional powerhouse** that balances **photo excellence and video innovation** in a compact mirrorless body.

Here's a taste of what makes it special:

- **High-Resolution 45+ Megapixel Sensor** – Capture detail-rich images that are print-worthy even at large sizes.

- **Next-Gen Autofocus** – Powered by deep learning, the R5 Mark II's autofocus tracks eyes, animals, vehicles, and more — fast and reliably.

- **Mind-Blowing Video** – Shoot in 8K RAW or 4K at high frame rates, with Canon Log and other pro-level tools at your fingertips.

- **Advanced Image Stabilization** – Combine in-body and lens IS for gimbal-smooth handheld shots.

- **Dual Card Slots** – Designed for professional workflows and high-volume shooting.

- **Weather Sealing & Build** – Tough enough for fieldwork, refined enough for the studio.

But features alone don't make great photos or videos, you do. That's where this book comes in.

How to Use This Book: Learn, Apply, Shoot

This guide is designed to do more than teach, it's built to help you act, improve, and grow.

Each chapter will:

- Explain the tools and techniques in plain language.

- Show you real-world applications, when and why you'd use a feature.

- Give you actionable tips, assignments, or checklists so you can practice what you've learned.

- Build your confidence step-by-step, from "Which button is which?" to "I'm confidently shooting a pro-level portrait session."

Feel free to move through the chapters at your own pace. Want to jump straight to advanced video recording or autofocus settings? Go for it. Prefer to master the basics first? That's perfect too.

Think of this book as your friendly instructor in your camera bag, always ready to guide you, explain things simply, and help you get the shot.

This front-facing image of the Canon EOS R5 Mark II showcases its sleek design and professional build, making it an ideal visual introduction for readers new to the camera. It highlights the camera's ergonomic layout and gives a clear view of its controls, setting the stage for the

detailed exploration in the subsequent chapters.

Chapter 2

Unboxing & Setup

So, your Canon EOS R5 Mark II has finally arrived exciting, isn't it? Before you jump into capturing breathtaking shots, let's start with the basics: unboxing, getting everything set up, and preparing the camera for its first real use.

What's in the Box

When you open that Canon-branded box, you should find the following items:

✅ **Canon EOS R5 Mark II Camera Body**

✅ **Canon LP-E6NH Rechargeable Lithium-Ion Battery**

✅ **Battery Charger LC-E6**

✅ **Camera Strap (with Canon R branding)**

✅ **Body Cap (to protect the sensor when no lens is attached)**

✅ **User Manuals / Warranty Card / Quick Start Guide**

✅ **Interface Cable (USB-C)**

⚠ **Note:** If you purchased a kit version, it may include a Canon RF lens (e.g., RF 24-105mm). Otherwise, lenses are sold separately.

Before going further, make sure all components are present. If something's missing or damaged, contact your seller right away.

Charging and Inserting the Battery & Memory Cards

🔋 **Step 1: Charging the Battery**

- Plug the charger into a wall outlet.
- Insert the LP-E6NH battery into the charger.

- The orange light will blink while charging and stay solid when fully charged (usually takes about 2–3 hours for a full charge).

Tip: Having a spare battery is a smart idea, especially for long shoots.

Step 2: Inserting the Battery

- Open the battery compartment at the bottom of the camera.
- Insert the battery, matching the metal contacts.
- Close the compartment securely.

Step 3: Inserting Memory Cards

- Open the memory card door on the right side of the camera (grip side).
- The R5 Mark II supports dual card slots:

   - CFexpress Type B (Slot 1) – ideal for 8K video and high-speed burst shooting.

- **SD UHS-II Card (Slot 2)** – for general photos and standard video use.

- Insert the cards label-side toward you, gently pushing until they click into place.

Use high-speed cards for best performance, especially if you're shooting video or burst mode.

First-Time Setup

Now that your camera is powered and ready, it's time to walk through the initial settings. When you power it on for the first time, the camera will guide you through some essential steps.

1. Set the Date, Time, and Time Zone

- This helps organize your photos and videos by accurate timestamps.

- Use the touchscreen or dial to make adjustments.

2. Choose Your Language

- Pick your preferred language (English, French, Spanish, etc.) for the camera menus.

3. Format Your Memory Card

- Go to Menu > Wrench Icon > Format Card.
- Select each card slot and format them before use.
- This erases any previous data and prepares the card for the R5 Mark II's file system.

Always format in-camera, not on your computer, to avoid corruption issues.

4. Register the Camera (Optional but Recommended)

- You can register the R5 Mark II on Canon's website using the serial number.
- This helps with warranty support and firmware update notifications.

✓ You're Ready!

Your Canon EOS R5 Mark II is now ready to roll, batteries charged, memory inserted, and menus set. You're officially one step closer to mastering one of the most powerful hybrid cameras on the market.

In the next chapter, we'll take a tour of the camera's body learning the buttons, dials, touchscreen, and key ports so you can handle it confidently.

Chapter 3

Camera Anatomy Made Simple

Before you dive into the fun stuff, shooting, filming, and creating, you need to get familiar with the tool in your hands. The Canon EOS R5 Mark II is packed with buttons, dials, and ports, but don't worry, by the end of this chapter, you'll know exactly what each part does and when to use it.

Think of this as your camera tour, where you meet the controls that will become second nature in no time.

1. The Front of the Camera

- Let's start with what you see when you look at the front of the R5 Mark II:

- **Lens Mount** – This is where you attach your RF lenses. It's large and sturdy for a reason: more light, more quality.

- **Lens Release Button** – Push this to remove your lens.

- **AF Assist Beam Emitter** – Helps the camera focus in dark environments.

- **Shutter Button** (seen slightly from the top) – Where you take your photos.

- **Front Dial** – Used for changing settings like aperture or shutter speed.

- **Customizable Function Button (Fn)** – You can assign this to whatever feature you use often.

Tip: Try pressing the shutter halfway to activate autofocus, that'll become muscle memory fast!

2. The Back of the Camera

- The rear side is where most of the action happens:

- **3.2" Vari-Angle Touchscreen LCD** – Your visual hub. Tap to focus, swipe to review photos, or use it for full menu navigation.

- **Multi-Controller Joystick** – Navigate menus or move the focus point with your thumb.

- **AF-ON Button** – Preferred by many pros for "back-button focusing."

- **Main Control Dial** – Adjust settings on the fly.

- **Q (Quick Menu) Button** – Opens a customizable shortcut menu.

- **Menu Button** – Takes you to the full settings interface.

- **Playback & Delete Buttons** – Review and manage your shots.

The AF-ON and joystick combo is especially useful for sports, wildlife, and fast-action shooting.

3. The Top of the Camera

Looking down from above, you'll see:

- **Top LCD Display** – Quick-glance info about your settings (aperture, shutter speed, ISO, battery life, etc.)

- **Mode Dial** – Switch between Manual, Aperture Priority, Shutter Priority, Auto, C1–C3 custom modes, etc.

- **Multi-Function (M-Fn) Button** – Great for toggling focus modes or exposure settings.

- **Record Button** – Instantly start/stop video recording.

- **Hot Shoe Mount** – Attach flashes, microphones, or other accessories.

You can switch seamlessly between photo and video modes using the mode dial and the dedicated Record button.

4. The Sides of the Camera

- The sides house your ports and card slots:

- Left Side (when looking from behind):

- Microphone Input (3.5mm)

- Headphone Output

- Remote Control Terminal

- **USB-C Port** – For charging, tethering, and file transfers.

- **Micro-HDMI Port** – For connecting external monitors or recorders.

Right Side:

- Dual Card Slot Door

  - **Slot 1:** CFexpress Type B

  - **Slot 2:** SD UHS-II

If you're shooting 8K video or fast bursts, use Slot 1 (CFexpress). For everyday stills, Slot 2 (SD) works great.

✅ Recap

You're now familiar with every important dial, button, port, and screen on your Canon EOS R5 Mark II. This foundational knowledge will help you move faster and more confidently when adjusting settings or troubleshooting.

Part Names

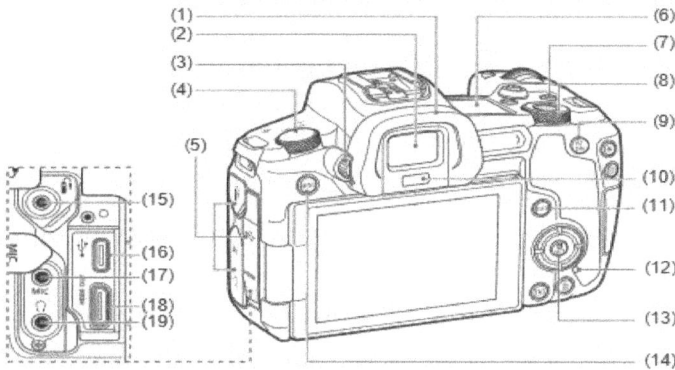

(1)	Eyecup	(11)	<INFO> Info button
(2)	Viewfinder eyepiece	(12)	Access lamp
(3)	Dioptric adjustment knob	(13)	<⊡/SET> Quick Control/Setting
(4)	Power switch		button
(5)	Terminal cover	(14)	<MENU> Menu button
(6)	LCD panel	(15)	<⋒> Remote control terminal
(7)	<LOCK> Multi-function lock button	(16)	<⟷> Digital terminal
(8)	<⟐> Quick control dial	(17)	<MIC> External microphone IN
(9)	<⚏> AF start button		terminal
(10)	Viewfinder sensor	(18)	<HDMI OUT> HDMI mini OUT
			terminal
		(19)	<⌒> Headphone terminal

31

Chapter 4

Menu System Demystified

When it comes to mastering your camera, knowing how to navigate the menu system is essential. The Canon EOS R5 Mark II has a wealth of options to help you customize the camera to your exact needs. But don't worry, we'll take it step by step, starting with the basics and moving into more advanced functions.

By the end of this chapter, you'll not only understand how to navigate the menus but also learn how to set up your camera for quick, efficient use.

1. Menu Walkthrough: Easy Mode vs. Advanced Mode

The Canon EOS R5 Mark II provides two primary navigation options: Easy Mode and Advanced Mode. Let's look at both:

Easy Mode (Beginner-Friendly)

When you first use the camera, you'll probably want to start with Easy Mode. This simplifies the interface, showing only essential functions.

- **Icons:** The options are displayed as easy-to-understand icons, not lists.

- **Simplified Settings:** Only the most frequently used features (like ISO, exposure, etc.) are visible.

- **Automatic Features:** Many settings are set to automatic, making it easier for beginners to take great photos without needing to know too much about the technical aspects.

Tip: If you're just starting out or want to take some shots without diving into settings, Easy Mode is your best friend.

Advanced Mode (Full Control)

As you get more comfortable with the R5 Mark II, you'll

want to switch to Advanced Mode to access deeper settings and control over your images.

- **Full Menu Access:** You'll see all available settings, from exposure controls to autofocus modes.
- **Customization:** Allows you to customize settings like picture styles, focus modes, and much more.
- **Detailed Information:** The Advanced Mode offers more detailed readouts of your current settings.

Advanced Mode is ideal for when you want to take full control of your camera, giving you the flexibility to shoot exactly how you want.

2. "My Menu" Setup for Quick Access

Now, let's make the camera even easier to use. If you find yourself constantly changing the same settings like ISO, white balance, or autofocus you can create a "My Menu". This is your personal shortcut to the most important settings, all in one place.

How to Set Up "My Menu"

- **Go to the Menu:** Press the Menu button.

- **Select the "My Menu" Tab:** It's usually located at the far right of the menu options.

- **Add Menu Items:** Navigate to the setting you use most (e.g., ISO, White Balance, Custom Functions) and add it to your "My Menu".

- **Save:** Hit Set to confirm the items you've selected.

Tip: You can add anything you use regularly to the "My Menu" tab. It's like creating your own personalized control panel.

3. Dummy-Friendly Menu Recommendations

As a beginner, it's easy to feel overwhelmed by the number of menu options available. Here are a few simple recommendations to help you navigate the menu without getting lost:

Quick Menu Setup

- **ISO:** Adjust the ISO for low light or fast shutter speeds. Keeping it in your "My Menu" means you can change it in a snap.

- **White Balance:** For accurate color. You can set this based on your environment (e.g., daylight, tungsten light).

- **Autofocus Settings:** Get familiar with AF area selection and focus modes. Having these at your fingertips will make focusing faster and more reliable.

- **Drive Mode:** For switching between single-shot, continuous shooting, or self-timer. Great for action shots or selfies.

Picture Style

- **Standard:** A good default for everyday shooting.
- **Portrait:** Enhances skin tones.

- **Landscape:** Deepens the blues and greens, perfect for nature shots.

- **Monochrome:** For black-and-white photography.

Pro Tip: If you're shooting landscapes, switch to Landscape Picture Style for more vivid colors.

Custom Functions

Once you feel more comfortable with the camera, experiment with custom functions to fine-tune your shooting experience. These include things like:

- **Custom Buttons:** Assign certain functions to the buttons you use most.

- **Focus Peaking:** Helps you focus manually by highlighting areas of contrast.

Don't worry about perfect settings right away. As you grow more confident, you can explore these advanced features.

✅ Menu Recap

You've now learned how to navigate the Canon EOS R5 Mark II's menus, set up quick access through "My Menu," and found key settings for everyday shooting. The more you practice with the menu, the more intuitive it will become.

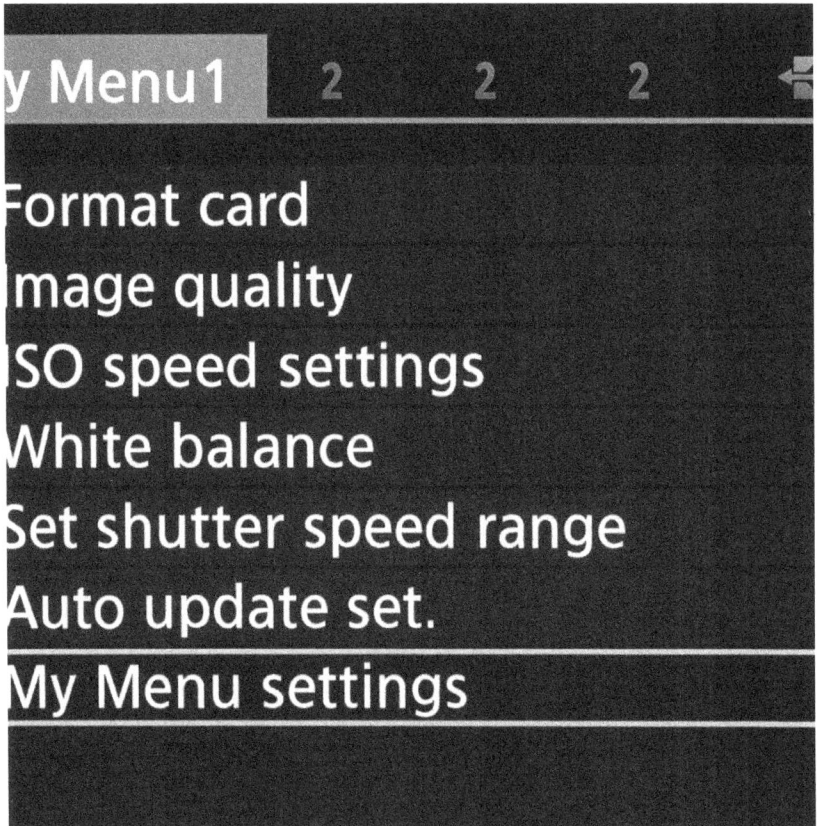

Chapter 5

Snap Your First Photos

Congratulations! You've unboxed your camera, set it up, and now understand the controls. It's time to take your first real photographs. This chapter is all about getting comfortable with the basics of shooting from how you hold the camera to snapping your first image, and finally how to review and manage what you've captured.

1. How to Hold and Stabilize the Camera

Even with top-tier tech, shaky hands can ruin a great shot. Let's start with the proper stance.

Proper Grip:

- **Right Hand:** Wrap your right hand firmly around the grip, with your index finger resting lightly on the shutter button.

- **Left Hand:** Cradle the lens from below with your left hand. This helps support the camera and allows easy control of zoom or focus rings.

- **Viewfinder vs. LCD:** For extra stability, tuck the camera close to your face and use the electronic viewfinder. This reduces camera shake.

Good Posture:

- **Feet Shoulder-Width Apart:** This balances your body.

- **Elbows In:** Tuck them close to your body, not out like wings.

- **Breath Control:** Exhale gently before pressing the shutter to avoid motion blur.

Bonus Tip: Use a neck strap for added security not just to protect from drops, but to help stabilize shots.

2. Auto Mode Walkthrough (Scene Intelligent Auto)

If you're new to photography, Auto Mode (green "A+"

icon on the mode dial) is your best friend. It does all the thinking for you, adjusting:

- Exposure

- Focus

- ISO

- White balance

To use Auto Mode:

- **Turn the Mode Dial** to the green "A+" (Scene Intelligent Auto).

- **Point the Camera** at your subject.

- **Half-Press the Shutter** Button to focus.

- **Fully Press the Shutter** to take the shot.

The camera analyzes the scene and chooses the best settings automatically. It's perfect for:

- Portraits

- Landscapes

- Action shots

- Low light scenes

Try shooting indoors and outdoors in Auto Mode to see how the camera adapts to lighting.

3. Reviewing, Deleting, and Saving Images

Once you've taken a photo, it's important to review and decide whether to keep or discard it. Here's how:

Review Your Photos

- Press the ▶ (Playback) button to see the last image taken.
- Use the multifunction dial or touch swipe to scroll through your shots.
- Zoom In/Out using the magnifying glass buttons or by pinching on the touchscreen.

Delete Unwanted Photos

- While reviewing an image, press the Trash Can/Delete button.

- Confirm the deletion when prompted.

Tip: Be careful! Deleting from the camera is permanent unless you've backed up.

Save and Transfer

Eventually, you'll want to save your photos to a computer or cloud:

- **Option 1:** Remove the memory card and insert it into a card reader.

- **Option 2:** Use a **USB-C** cable to connect your camera directly to your computer.

- **Option 3:** Enable **Wi-Fi/Bluetooth** in the menu to transfer files wirelessly to Canon's Camera Connect app.

Create a habit: Review after each shoot, delete only duplicates or mistakes, and back up your best shots!

✅ Ready to Shoot?

By the end of this chapter, you should be able to:

- Hold the camera with confidence

- Take photos in Auto Mode

- Review and manage your images with ease

PART 2: UNDERSTANDING PHOTOGRAPHY – LEARN TO SHOOT LIKE A PRO

Chapter 6

The Exposure Triangle (Dummy Style)

If photography were a video game, **ISO**, **shutter speed**, and **aperture** would be your three main powers. Together, they form the **Exposure Triangle** a balancing act that controls how light or dark your photos are. Mastering this triangle helps you take control of your shots instead of relying on Auto Mode.

But don't worry we'll break it down so simple, your grandma could shoot portraits by the end of this chapter.

▲ What Is the Exposure Triangle?

Imagine a triangle with these three elements at each point:

- **ISO** – Sensitivity to light
- **Shutter Speed** – How long the sensor is exposed
- **Aperture (f-stop)** – How wide the lens opens

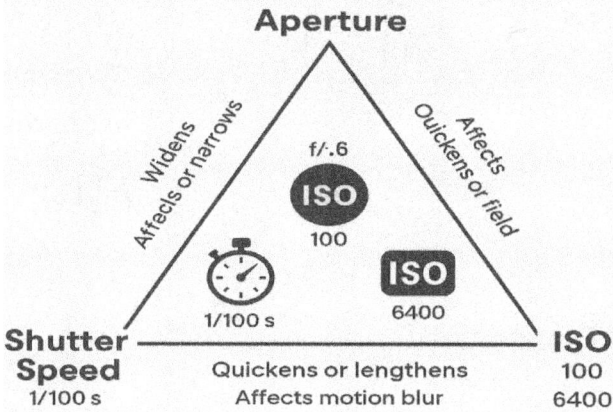

THE EXPOSURE TRIANGLE

Each affects exposure (brightness), but they also bring side effects:

Setting	What it Does	Side Effect
ISO	Brightens or darkens the photo	Too high = grainy/noisy image
Shutter Speed	Controls motion (freeze or blur)	Too slow = motion blur

Aperture	Controls depth of field	Wide open = blurry background (bokeh)

Visual Explanation

☼ ISO (Think: Sensitivity)

- **Low ISO (100–400):** Clean, low noise – best for bright conditions
- **High ISO (1600–12800+):** Good for low light – but introduces grain

◷ Shutter Speed (Think: Time)

- Fast (1/1000s): Freezes motion (great for sports)
- Slow (1/30s or longer): Shows motion blur (ideal for waterfalls or night shots)

☾ Aperture (Think: Opening)

- Wide (f/1.8): More light, shallow focus (blurry background)

- Narrow (f/16): Less light, deep focus (everything sharp)

Tip: The lower the f-number, the bigger the opening.

Practice Drills with Cheat Codes

Let's learn by doing. Try these mini drills:

Drill 1: Blurry Background Portrait

- Mode: Av (Aperture Priority)

- Set Aperture: f/2.8

- ISO: Auto

- Result: Person sharp, background blurred

Drill 2: Freeze Action

- Mode: Tv (Shutter Priority)

- Shutter: 1/1000s

- ISO: 800

- Result: Crisp motion freeze of jumping pet or child

Drill 3: Low Light Without Flash

- Mode: Manual

- ISO: 3200

- Aperture: f/1.8

- Shutter: 1/60s

- Result: Bright enough indoor shot, no flash needed

Dummy Cheat Codes Summary

Goal	Cheat Code
Blur background	f/1.8 or f/2.8 (Av mode)
Freeze action	1/1000s shutter (Tv mode)
Avoid grain	Keep ISO low (100–400)

| Low light, no flash | High ISO + wide aperture (M mode) |

✅ By the End of This Chapter

You now:

- Understand ISO, aperture, and shutter speed
- Know how to balance them using triangle logic
- Can start experimenting in manual or semi-auto modes

Chapter 7

Getting Focus Right

Blurry shots are a photographer's nightmare but the Canon EOS R5 Mark II comes with cutting-edge autofocus (AF) that can turn you into a sharp-shooting pro, even if you're just starting out. This chapter makes the camera's powerful AF system feel like second nature.

Autofocus Basics: What is AF?

Autofocus (AF) is how your camera locks onto a subject without needing you to twist the focus ring manually. The R5 Mark II is especially advanced, with AI-powered **Eye/Face tracking**, **animal detection**, and **subject recognition**.

Let's break it down into dummy-friendly focus modes first.

One-Shot vs Servo (AF)

These are the two main AF modes you'll use the most:

Mode	Best For	What It Does
One-Shot AF	Still subjects (portraits, objects)	Focuses once when you half-press the shutter
Servo AF	Moving subjects (kids, pets, sports)	Continuously tracks and adjusts focus

Tip: Use One-Shot for calm scenes. Use Servo when there's motion.

👁 Eye/Face Tracking (Simplified)

The R5 Mark II can recognize and follow human eyes, faces, and even animals. When enabled:

- **Eye Detection AF:** Locks precisely on the closest visible eye.

- **Face Tracking AF:** Tracks a face even as it moves across the frame.

- **Subject Detection:** Identifies animals, birds, or vehicles depending on your selected mode.

How to enable:

- Go to **AF Menu** > Subject Detection
- Choose: **People**, **Animals**, or **Vehicles**
- Turn on **Eye Detection AF**

Use this when shooting portraits, weddings, or wildlife you'll be amazed how "smart" the focus is.

Practice Drill: Focus on a Moving Subject

Goal: Get a sharp shot of a person walking toward you.

Setup:

1. Set AF mode to **Servo**

2. Enable **Eye Detection**

3. Use **Continuous shooting** mode

4. Track the person walking with the shutter half-pressed or using **Back-Button Focus**

Result: The camera will keep the subject's eye in sharp focus as they move.

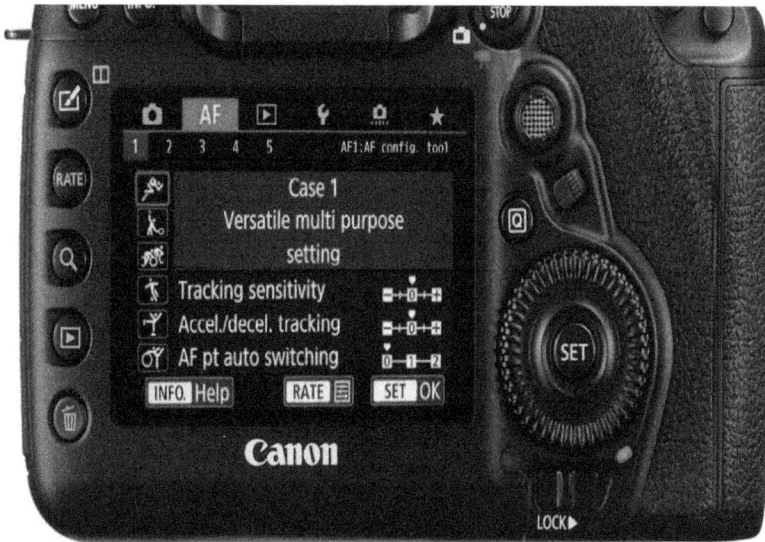

Focus Cheat Sheet Summary

Need	Use
Portrait (still)	One-Shot + Eye Detection
Running pet or child	Servo AF + Face/Eye Tracking
Birds in flight	Servo + Subject = Animal/Bird
Static objects (products, food)	One-Shot AF

✅ By the End of This Chapter

You'll know:

- The difference between One-Shot and Servo
- How to activate and trust Eye/Face tracking
- How to practice real-world focus techniques

Chapter 8

JPEG vs RAW – Which Should You Use?

One of the first choices every photographer faces is whether to shoot in JPEG or RAW. Don't worry you're not alone if you've wondered what these formats even are. This chapter will break it down dummy-style, so you'll never feel confused again.

What is JPEG?

JPEG (or JPG) is a **compressed image format**. It's the camera's way of saying:

"Let me process this photo for you, shrink the file, and make it ready to use."

✅ Pros of JPEG:

- Smaller file sizes = more photos per memory card

- Instantly viewable and shareable

- Built-in sharpening, contrast, and color

- Great for beginners and quick social media use

✖ Cons of JPEG:

- Limited editing flexibility

- Some image data is permanently lost

- Not ideal for serious post-processing

What is RAW?

RAW is like a digital **film negative**. It captures **all the data** from the sensor with minimal in-camera processing.

It's the purest form of your photo, meant to be edited later on your computer.

✅ Pros of RAW:

- Maximum detail and dynamic range

- Excellent for editing exposure, color, shadows, etc.

- No compression = best image quality

✖ Cons of RAW:

- Large file sizes = fewer photos per card

- Needs editing software (like Adobe Lightroom)

- Not directly shareable without conversion

When Should You Use Each?

Situation	Best Format
Quick snapshots, social media	JPEG
Events where storage is limited	JPEG
Professional shoots, portraits	RAW
High-contrast lighting scenes	RAW

You plan to edit your RAW

photos

Tip: If you're unsure, many cameras (including the R5 Mark II) let you shoot both at once: RAW + JPEG.

Storage & Editing Implications

Format	Average File Size	Needs Editing?	Easy to Share?
JPEG	3–10 MB	No	Yes
RAW	25–50 MB	Yes	No (must export first)

- **Storage Tip:** Get fast UHS-II or CFexpress cards if shooting RAW often
- **Editing Tip:** Use software like Lightroom, Canon Digital Photo Professional, or Capture One

Dummy Summary: RAW vs JPEG Cheat Code

If You Want...	Choose...
Fast sharing, no editing	JPEG
Best quality for pro results	RAW
A balance between the two	RAW + JPEG

✅ By the End of This Chapter

You now understand:

- What JPEG and RAW really are
- The pros and cons of each
- How to choose based on your needs

Chapter 9

Color, Light & White Balance

If your photos ever look too yellow, too blue, or just "off," chances are the lighting and white balance weren't quite right. This chapter makes it simple to understand how light and color shape your images and how to control them like a pro.

Basic Lighting Tips

Light is everything in photography. It sets the mood, reveals detail, and affects sharpness and color. The R5 Mark II's sensor loves good light so let's learn how to feed it properly.

Three Main Types of Light:

Type	Source Example	Tip
Natural Light	Sunlight, skylight	Use during golden hour for soft tones
Artificial Light	Lamps, bulbs, LEDs	Watch for color casts (orange/green)
Flash Light	On-camera or external flash	Use to fill shadows or freeze action

Pro tip: Place your subject near a window for soft, flattering light.

Auto White Balance (AWB) vs Manual

White Balance (WB) adjusts the colors in your photo so that whites look white not bluish or orange.

Auto White Balance (AWB)

- The camera guesses based on the scene.
- Works well in most situations.

- Can struggle under mixed lighting (e.g., indoors with window light + bulbs).

Manual White Balance

- Lets you set specific color temperatures.
- Choose from presets: Daylight, Shade, Cloudy, Tungsten, Fluorescent, etc.
- Custom WB: Use a gray card or white surface to fine-tune color accuracy.

Tip: Try "Kelvin Mode" for full control — set color temperature manually (e.g., 5200K for daylight).

Using Natural vs Artificial Light

Scenario	Recommended Light	Why?
Outdoor portraits	Natural (shade)	Softer skin tones, flattering shadows

Indoor room photos	Artificial (LED)	Consistent lighting
Product photography	Mix of both	Use daylight + lightbox or soft LED
Events or weddings	Flash + ambient	Fill dark spots, freeze moments

Mixed lighting? Stick to one dominant source or use gels to match color.

Quick Practice Drill

1. Take the same subject in:

- Natural light (window)
- Warm indoor light (lamp)
- Mixed light (both)

2. Use AWB, then switch to preset WB, then try Kelvin

Compare the results this exercise trains your eye fast.

Dummy-Friendly Summary

Goal	White Balance Mode
General shooting	AWB
Indoor with yellow light	Tungsten or Kelvin 3200K
Outdoors on cloudy day	Cloudy or Kelvin 6000K
Studio or flash setups	Flash or Custom WB

WHITE BALANCE

ıto	Daylight	Shade	Cloudy	Tungsten	White	Flasl
ıera	neutral	warm	slightly	cool	fluorescent	neutrɛ
ɔses			warm			

COLOR TONONE

✅ By the End of This Chapter

You've learned:

- The difference between types of light

- How to fix strange color casts

- How to use white balance for accurate, vibrant photos

PART 3: UNLOCKING MANUAL CONTROL (NO MORE AUTO!)

Chapter 10

Taking Control – Manual Modes

Ready to graduate from Auto mode? This chapter will demystify the four key camera modes P, Tv, Av, and M and explain when and why to use each. No jargon, no fluff, just clarity.

Canon's Big Four Modes (P, Tv, Av, M)

Let's break them down with a simple cheat sheet:

Mode	Stands For	What You Control	Camera Controls	Best For
P	Program AE	Exposure Compensation	Shutter & Aperture (auto combo)	Beginners stepping out of Auto
Tv	Shutter Priority	Shutter Speed	Aperture	Motion control (action, sports)
Av	Aperture Priority	Aperture (f-stop)	Shutter Speed	Portraits, depth control

M	Manual	Everything (ISO, SS, Aperture)	Nothing auto	Full control & creative shooting

---

Tip: "P" is Auto with options. "M" is the pilot seat. "Tv" = Time (shutter). "Av" = Aperture (background blur).

Scenario-Based Examples

1. Capturing Fast Action – Use Tv Mode

- Set shutter speed to 1/1000s
- Camera picks matching aperture
- Perfect for: sports, wildlife, kids running

2. Beautiful Background Blur – Use Av Mode

- Set aperture to f/1.8 – f/2.8
- Camera picks shutter speed
- Perfect for: portraits, food, close-ups

3. Night Landscape – Use M Mode

- Set shutter: 15s, aperture: f/8, ISO: 100

- Use tripod!

- Perfect for: cityscapes, light trails

4. Walk-around Daylight Shooting – Use P Mode

- Camera picks combo, but you can tweak ISO or exposure compensation

- Great for: quick snapshots when you don't want to miss a moment

Switching Modes on the R5 Mark II

- Turn the Mode Dial on top of the camera.

- Watch the LCD change to match your selection.

- Use the Main Dial or Quick Menu (Q) to adjust exposure settings easily.

Quick Visual Metaphor

- **P Mode** = Cruise control (safe but flexible)

- **Tv Mode** = Freezing motion or showing blur

- **Av Mode** = Background blur for storytelling

- **M Mode** = Full creative command

P

YOU CONTROL	CAMERA CONTROLS
Exposure Compensa-tion	Shutter Speed

Tv

YOU CONTROL	CAMERA CONTROLS
Shutter Speed	Aperture

Av

YOU CONTROL	CAMERA CONTROLS
Aperture	Shutter Speed

M

YOU CONTROL	CAMERA CONTROLS
ISO· Shutter Speed- Aperture	None

If the lighting is changing fast, Av or Tv will help you stay adaptable while still making key creative choices.

✅ Summary Table for Quick Reference

Scene Type	Recommended Mode	Why
Sports or action	Tv	Freeze or blur motion
Portraits with bokeh	Av	Control depth of field
Artistic control	M	Adjust all settings manually
General street/travel	P	Let camera assist, tweak as needed

Chapter 11

Using the Histogram & Exposure Tools

Ever taken a photo that looked great on the LCD, but later realized it was too dark or blown out? That's where the histogram and exposure tools come to the rescue.

📈 What Is a Histogram?

A histogram is a graph that shows the brightness levels in your photo from black to white.

Visual Breakdown:

- **Left side** = Shadows (blacks)
- **Middle** = Midtones (grays)
- **Right side** = Highlights (whites)

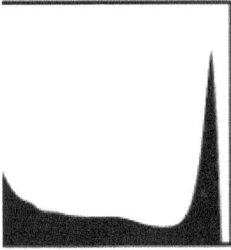

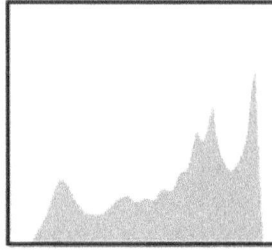

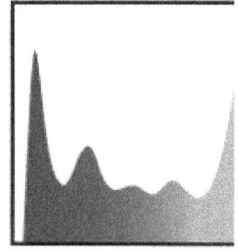

)w Slim'tons Haihts	Shadows Midtones Whe	Hidtones Heighaxt
DEREXPOSED	OVEREXPOSED	WELL-EXPOS

It doesn't show the picture, but how the light is spread across it.

How to Read a Histogram (The Dummy Way)

Histogram Shape	What It Means	Fix Suggestion
Bunched to the left	Too dark (underexposed)	Increase exposure / raise ISO
Bunched to the	Too bright	Lower exposure /

right	(overexposed)	use faster shutter
Spread evenly across	Balanced exposure	Usually a good result
Tall spike on one side	Lots of detail in that tone (dark/light)	May be - OK depends on subject

Think of it like a sound equalizer, you want a good balance unless your style calls for extremes.

How to Use It on the R5 Mark II

1. Enable Histogram Display:

- In *Playback*: Press [INFO] until the histogram appears.

- In *Live View*: Use [INFO] button while composing a shot.

2. Check Before & After You Shoot

- Use *Live View Histogram* for real-time exposure aid.
- Review image histogram after capture to confirm accuracy.

3. Types of Histograms

- *Brightness (Luminance)* – most useful
- *RGB Histogram* – shows color clipping (for pros)

Exposure Compensation in Real Life

Use this when your photos look too dark or too bright especially in Program, Tv, or Av mode.

How:

- Look for the +/- exposure scale on the LCD.
- Turn the top dial (next to the shutter) to shift exposure:

  - Move right (+) = brighter

- Move left (−) = darker

📷 *Shooting in snow? Dial in +1.0 EV. Shooting indoors? Try -0.3 EV to preserve highlights.*

Practice Drill

1. Take a photo in Av mode of a well-lit scene.

2. Review the histogram:

- Is it leaning to one side?

- Adjust exposure compensation and try again.

3. Compare results and histogram changes.

✅ **Quick Summary**

Tool	What It Does	Use When...
Histogram	Shows light balance in your image	You want to confirm correct exposure
Exposure	Adjusts image	Your shots look

Compensation	brightness manually	too dark or too bright
Live View Histogram	Helps you correct before shooting	You want to "expose to the right" or avoid clipping

Chapter 12

Autofocus Like a Ninja

A fast subject. Low light. An animal running. You need your Canon R5 Mark II's Autofocus (AF) to work like a ninja: fast, precise, silent. Let's simplify the complex AF system into something you can actually use.

AF Area Modes Simplified

The R5 Mark II offers multiple AF area options choosing the right one is key.

AF Area Mode	What It Does	Best For
Spot AF	Focuses on a tiny pinpoint	Macro shots, high precision

1-Point AF	Small focus box, manually movable	General photography, still subjects
Expand AF Area	1-Point + surrounding sensors for support	Subjects that may shift slightly
Zone AF	Focuses within a selectable rectangular zone	Moving subjects in a specific area
Large Zone AF (Vertical/Horizontal)	Wider area than Zone AF	Sports, animals, fast motion
Whole Area AF	Camera picks subject	People, events, fast-changing

	automatically	scenes
Eye/Face/Head Detection	Tracks human faces and eyes precisely	Portraits, vlogging, interviews
Animal/Bird/Vehicle AF	AI-powered subject detection	Wildlife, pets, motorsports

Tip: Think "smaller box = more precision," "larger zone = more flexibility."

Tracking People, Animals & Fast Subjects

Canon's Deep Learning AF lets you track nearly anything:

🧍 For People:

- Use Eye **AF** + **Whole Area**
- Works even if subject turns or wears glasses

🐾 For Animals:

- Activate **Animal Detection AF**

- Works with dogs, cats, birds even some exotic species

⚓ For Vehicles:

- Turn on **Vehicle Detection**

- Recognizes cars, bikes, even helmets

🎥 *In video mode, Eye AF and subject tracking remain incredibly sticky even when they move quickly across the frame.*

Low-Light Focusing Tips

Autofocus can struggle in the dark, here's how to help:

- **Use a faster lens** (f/1.8, f/2.8) — lets in more light

- **Switch to 1-Point AF** — helps focus on a specific contrast area

- **Use focus assist beam** — built into Canon flash units

- **Try Manual Focus if AF fails** — use focus peaking in Live View

Bonus tip: Aim at something with contrast (like a window frame), not a flat wall.

Quick Setup Walkthrough

1. Press the [AF] button or tap the AF icon on the touchscreen.

2. Choose your **AF method** (Whole Area, Eye Detect, etc.). Use the **multi-controller joystick** to move focus box if needed.

3. Half-press shutter to confirm tracking is locking correctly.

Practice Drill

Track a walking person across your yard.

Use **Whole Area** + **Eye Detection AF**.

Observe how the focus follows even when they turn sideways.

Repeat at night using **1-Point AF** and a porch light.

✅ Summary

Situation	Best AF Mode
Portraits	Eye Detect + 1-Point or Whole Area
Pets/Animals	Animal Detection + Whole Area
Fast Action	Large Zone or Whole Area AF
Low Light	1-Point AF + fast lens
Macro/Detail Shots	Spot or 1-Point AF

Chapter 13

Better Composition Equals Better Photos

Even with the best camera, a poorly composed image can still feel... meh. But when you nail composition? Your photos instantly look sharper, more professional, and more engaging no fancy editing needed.

The Basics: 3 Composition Superpowers

1. Rule of Thirds

Imagine your frame divided into 9 equal parts — 2 vertical lines, 2 horizontal.

- Place your subject where those lines cross (the power points).

- Helps balance the image and guide the viewer's eye.

Tip: Turn on the "grid display" in your camera's viewfinder to help!

2. Symmetry & Patterns

- Humans *love* symmetry, it feels tidy and pleasing.
- Capture reflections, buildings, even faces centered.

But don't overdo it. Breaking symmetry can also create tension and interest.

3. Depth and Layers

Make your photos feel 3D by including:

- A **foreground** (something near the lens)
- A **midground** (your main subject)
- A **background** (context or scenery)

Think: "What's in front, what's behind, and how do they relate?"

Framing Tips & Mistakes to Avoid

👍 DO This	⊘ Avoid This
Use natural frames (windows, trees)	Cutting off heads, feet, or important elements
Leave breathing space (headroom)	Cropping too tightly or awkwardly
Shoot at eye level or lower	Always shooting from standing height
Use leading lines (roads, fences)	Distracting backgrounds or clutter

Pro trick: Tilt your camera slightly for dynamic angles, but don't overdo it.

🎁 Bonus: What Makes a "Pro-Looking" Photo?

Here are some habits that elevate your shots:

- **Intentional framing:** Every corner of the frame has a purpose.

- **Simplicity:** Avoid too many elements fighting for attention.

- **Light direction:** Use shadows and highlights for mood.

- **Editing polish:** Slight contrast, clarity, or color tweaks can seal the deal.

- **Emotion or story:** What's happening in the photo and why should we care?

Practice Challenge

Take one subject (your coffee mug, pet, or friend) and:

1. Shoot it using rule of thirds.

2. Shoot it dead-center with symmetry.

3. Shoot it using layers: place something in the foreground, focus on your subject in the midground, and include a background.

4. Review: Which photo tells the strongest story?

✅ Quick Recap

Technique	What It Does	Great For
Rule of Thirds	Balanced, pleasing compositions	Portraits, landscapes
Symmetry	Clean, satisfying images	Architecture, reflections
Depth	Adds realism and storytelling	Travel, food, street scenes
Framing	Focuses attention	Creative portraits, storytelling

PART 4: LENS, GEAR & GENRE-BASED PRACTICE

Chapter 14

Mastering the RF Mount & Lenses

You've got the powerful Canon EOS R5 Mark II in your hands now it's time to unlock its full potential with the right lens. Think of lenses as the eyes of your camera: different lenses see the world differently.

What Lenses Work with the R5 Mark II?

The R5 Mark II uses Canon's RF mount a modern lens system built for speed, sharpness, and advanced features.

Lens Type	Works With R5 Mark II?	Notes
RF Lenses	✓ Yes	Designed specifically for R-series cameras
EF Lenses	✓ With adapter	Use Canon's EF-RF adapter (no quality loss)
EF-S Lenses	✓ With adapter	Works, but image is cropped
Third-Party RF	✓ (Some)	Brands like Sigma, Tamron, Viltrox support RF mount

Stick with RF lenses when possible for the best autofocus and weather sealing.

Best Lenses by Genre

Here's a quick guide to choosing the right lens for what you shoot:

Genre	Recommended Lens	Why It's Great
Portraits	RF 85mm f/1.2L or RF 50mm f/1.8 STM	Creamy background blur, flattering compression
Travel	RF 24-105mm f/4L IS USM	Versatile zoom range, lightweight
Street	RF 35mm f/1.8 Macro IS STM	Small, fast, discreet
Vlogging	RF 15-30mm	Wide angle,

	f/4.5-6.3 IS STM	image stabilization
Wildlife/Sports	RF 100-500mm f/4.5-7.1L IS USM	Powerful zoom, fast AF
Macro	RF 100mm f/2.8L Macro IS USM	Life-size close-ups with stunning sharpness

On a budget? Try the RF 50mm f/1.8 — small, sharp, and under $200.

🔍 Prime vs Zoom Lenses (Simplified)

Feature	Prime Lens	Zoom Lens
Focal Length	Fixed (e.g., 50mm only)	Variable (e.g., 24-70mm)

Sharpness	Often sharper	Slightly less (but still excellent)
Size/Weight	Lighter and smaller	Heavier, more glass
Creativity	Forces you to move = better learning	More flexible
Best For	Portraits, low light, bokeh	Travel, events, general purpose

Pro Insight: Many photographers start with zooms, then fall in love with the simplicity of primes.

Quick Lens Tips

- Try before you buy: Rent a lens if you're unsure.
- Use the right tool: Don't shoot wildlife with a 35mm or portraits with a 16mm!

- Invest in glass: A good lens lasts longer than a camera body.

Practice Challenge

1. Pick a prime lens (like 35mm or 50mm).

2. Shoot a day in your life with just that lens.

3. Repeat using a zoom.

4. Compare: Which helped you see more creatively?

Chapter 15

Must-Have Accessories

Your Canon R5 Mark II is a powerhouse on its own but the right accessories can boost performance, simplify your shoots, and help you get professional results faster.

Let's break down the essentials every shooter should consider.

1. Memory Cards – Fast & Reliable Storage

Your R5 Mark II uses:

- **CFexpress Type B** (blazing fast, great for 8K/RAW video)
- **SD UHS-II** (more affordable, still very fast)

Use Case	Recommended Card

8K/4K Video	CFexpress Type B, 256GB+ (e.g., SanDisk Extreme Pro)
Photos + Occasional 4K	SD UHS-II, V90 rated (e.g., Lexar, ProGrade)

Always format new cards in-camera before use.

2. Tripods – Stability for Sharp Shots

A tripod is essential for:

- Long exposures
- Time-lapse or night photography
- Self-portraits and video work

Recommended:

- Manfrotto BeFree (portable)
- Peak Design Travel Tripod (premium & compact)

Tip: Choose one that folds small but feels sturdy.

3. Microphones – Better Audio for Video Creators

The R5 Mark II has great image quality, but audio? It can use some help.

Mic Type	Best For	Example
Shotgun	Vlogging, interviews	Rode VideoMic Pro+, Deity D4 Duo
Lavalier (clip-on)	Talking head videos	Rode Wireless GO II
Stereo Mic	Ambient sound capture	Zoom M3 MicTrak, Canon DM-E100

⚡ All plug into the 3.5mm jack or hot shoe mount.

4. ND Filters – For Smooth, Cinematic Footage

ND (Neutral Density) filters are like sunglasses for your lens.

They reduce light so you can shoot:

- Wide open (f/1.2, f/2.8) in bright sun

- 1/50 shutter speed for natural video motion

Types:

- **Fixed ND:** Specific stops (e.g., ND8, ND64)

- **Variable ND:** Adjustable strength, more flexible

Essential for video shooters using manual exposure.

5. Flashes & Lighting – Light Like a Pro

The R5 Mark II has no built-in flash, so consider:

- **Canon Speedlite EL-100 or 430EX III-RT** for on-camera

- **Godox V1-R** for powerful, soft, off-camera light

Bonus: Add a small LED panel (like Aputure AL-M9 or Lume Cube) for video lighting or portraits on the go.

6. Straps, Batteries, and Extras

Accessory	Why You'll Love It
Extra Batteries	R5 drains power fast (use LP-E6NH)
Battery Grip	Double battery life + better portrait grip
Neck/Shoulder Strap	Use padded or cross-body style
Lens Cleaning Kit	Keep glass spotless and smudge-free

Don't forget: A microfiber cloth, air blower, and lens pen go a long way.

✅ Quick Recap: Beginner's Kit Checklist

✅ SD UHS-II or CFexpress cards

✓ Lightweight tripod

✓ External mic (if doing video)

✓ ND filter (for outdoor video)

✓ Flash or LED light

✓ At least 2 extra batteries

✓ Camera strap and cleaning kit

PART 5: GOING PRO – ADVANCED TECHNIQUES & WORKFLOW

Chapter 16

Custom Modes & Button Mapping

Want to work faster, smarter, and more consistently? It's time to make your Canon R5 Mark II your camera.

This chapter will show you how to save your favorite setups (C1, C2, C3) and assign useful functions to buttons so you're always ready for the moment.

What Are Custom Modes (C1, C2, C3)?

Think of C1, C2, and C3 as memory banks for your favorite setups. You can save all your camera settings, shooting mode, ISO, AF setup, white balance, and more into one of these spots for instant recall.

✦ How to Set Up C1 / C2 / C3

1. Set the camera how you like it

- Choose your desired shooting mode (M, Av, etc.), ISO, focus settings, image quality, etc.

2. Go to Menu > Wrench Tab > Custom shooting mode (C1-C3)

- Select Register settings.

3. Choose C1, C2, or C3

- Save your current settings to the selected slot.

✅ That's it! Now you can turn the Mode Dial to C1/C2/C3 anytime and shoot instantly with your saved setup.

Sample Use Cases

Mode	Use Case	What It Stores
C1	Portraits	Av mode, Eye AF, f/2.8, Auto ISO, Single Shot AF
C2	Action/Wildlife	Manual, 1/2000s, Servo AF, High-Speed Burst
C3	Video Vlogging	Movie mode, Face Detect, 4K60p, Mic input active

Tip: You can re-register anytime if your needs change.

Button Customization: Your Shortcuts Panel

You don't need to dig into menus every time. Assign your

favorite tools to physical buttons.

How to Customize Buttons:

- Go to **Menu > Custom Functions (Orange Camera icon tab) > Customize buttons**

- Choose a button (e.g., AF-ON, SET, M-Fn)

- Pick a function from the list (like Eye AF, ISO, Movie Rec, etc.)

Smart Button Ideas

Button	Assign This To...	Why It Helps
AF-ON	Eye/Face AF Toggle	Instantly activate Eye AF
SET	ISO change or White Balance	Quick exposure changes
M-Fn	Drive Mode or AF Method	Switch between burst modes fast

Lens Control Ring	Aperture, ISO, or Manual Focus	On-the-fly tweaks while shooting

Bonus: You can also assign Touch & Drag AF, AF Method, or Zebra Display (for video) to custom buttons.

🚀 Speed Up Your Process

- Save your custom setups in My Menu for even faster access (covered in Chapter 4)
- Use Back-Button Focus (assign AF-ON for focus, shutter button only for exposure)
- Streamline your favorite genres with C1/C2/C3, so you're never caught adjusting settings in a rush

Chapter 17

Video Like a Pro

The Canon EOS R5 Mark II isn't just a photo beast it's a cinema-grade video camera in disguise. Whether you're shooting 4K reels or an 8K short film, this chapter will guide you from "press record" to pro-level confidence.

Video Resolutions & Frame Rates: What to Use and When

Setting	When to Use
4K 24/30p	Cinematic storytelling
4K 60p	Smooth motion, talking heads, YouTube
4K 120p	Super slow motion B-roll

8K 30p	Ultra-detailed scenes, futureproofing
1080p (Full HD)	Smaller file sizes, basic social content

Tip: 4K 60p strikes the best balance for most creators.

Video Formats Explained

- **MP4 (IPB):** Smaller file size, easier editing
- **ALL-I:** Higher quality, better for color grading
- **Canon Log (C-Log):** Flat color profile, more dynamic range ideal for post-production

🎨 Use C-Log3 if you want to color grade like a pro but know it needs editing!

Best Video Settings for Creators

Setting	Ideal Value
Shutter Speed	Double your frame rate (e.g. 1/60s for

	30p)
Aperture	f/2.8–f/5.6 for nice depth
ISO	Lowest clean value possible (ISO 100–800)
White Balance	Manual or custom for consistent color
Picture Style	Neutral or C-Log3 for editing
AF Method	Face + Tracking (AF Tab)

Better Audio = Better Video

- Use an external microphone (shotgun or lav mic)
- Monitor audio using headphones
- Set audio level manually: avoid Auto Gain pumping

Stabilization Tips

- In-Body IS + Digital IS = smoother handheld footage

- Use a tripod, monopod, or gimbal for best results

- Lens IS (if available) adds another layer of steadiness

Digital IS crops the frame slightly   keep that in mind for tight shots.

Sample Setup: YouTube Vlogging

- Mode: Movie Manual (M)

- Res/FPS: 4K 30p

- Lens: RF 16mm f/2.8 or 24-70mm f/2.8L

- Audio: Shotgun mic + windscreen

- Settings: Shutter 1/60, Aperture f/2.8, ISO Auto

- Stabilization: On

PART 6: MAINTENANCE, TROUBLESHOOTING & MASTERY

Chapter 18

Maintenance & Protection

Your Canon EOS R5 Mark II is a high-precision machine. A little care goes a long way in protecting your investment whether you're deep in the jungle or just shooting around town.

Cleaning Lenses & Sensor (Don't Panic)

Cleaning the Lens

- Front/Rear Elements:
- Use a blower first to remove dust.
- Gently wipe with a microfiber cloth or lens tissue.
- For smudges, use lens cleaning solution in small amounts.

Avoid circular polishing, wipe in straight lines from center out.

Cleaning the Sensor (Safely!)

- **Dust Spots on Images?** Try this first:

    - Menu > **Sensor Cleaning > Clean Now**

- Still there?

- Use a **manual air blower** (never canned air!)

- If necessary, use **sensor swabs** with sensor-safe fluid or better yet:

- **Take it to a Canon service center** for a deep clean.

Don't touch the sensor unless you're experienced — it's delicate.

Firmware Updates

Canon occasionally releases firmware to fix bugs, improve performance, or add features.

How to Update:

1. Go to Canon's support website

2. Download the latest firmware for the EOS R5 Mark II

3. Copy the **.FIR** file to a formatted memory card

4. Insert the card into the camera

5. Go to Menu > Firmware > **Update Firmware**

📷 *Make sure your battery is fully charged before updating.*

Travel Protection Tips

- **Use a padded camera** bag (with dividers)

- **Lens caps on at all times** during transport

- Bring a **rain cover or plastic bag** for sudden weather changes

- Store **silica gel packs** to prevent moisture build-up

- Use a **UV filter** or **lens protector** as a front shield

Packing Checklist

✓ Camera body + extra batteries

✓ Chargers + cables

✓ Lens cleaning kit

✓ Memory cards + case

✓ Rain cover + microfiber cloths

Chapter 19

Fix It Fast – Common Issues

Even the best camera can run into a hiccup now and then. This chapter is your emergency guide for fixing the most common problems quickly and confidently.

Problem: Blurry or Missed Focus

Symptoms: Subjects out of focus, Eye AF not locking, camera hunting in low light.

Quick Fixes:

- Make sure you're using the right AF mode:
- Still subjects → One-Shot AF
- Moving subjects → Servo AF
- Switch to Face/Eye Detection in menu (great for portraits)

- Tap the screen to refocus or half-press shutter to re-lock
- Use center point AF in tricky light
- Try brighter lighting or higher contrast targets

Tip: Clean your lens front element smudges can affect focus too.

Problem: Short Battery Life

Symptoms: Battery dies too fast, especially during video or cold weather.

Quick Fixes:

- Turn off Wi-Fi/Bluetooth when not in use
- Lower LCD screen brightness
- Use Power Saving Mode (Menu > Wrench tab > Eco Mode)
- Avoid continuous shooting or Live View for too long

- Carry extra LP-E6NH batteries

Note: Video (especially 8K) consumes much more power.

Problem: Camera Overheating

Symptoms: Camera shuts down after long video shoots or in hot environments.

Quick Fixes:

- Reduce recording resolution (drop from 8K to 4K HQ)
- Record in shorter clips, let the camera cool between takes
- Avoid direct sunlight; shade or fan the camera
- Use external recorders via HDMI if needed
- Remove the LCD from the body (open it slightly) to aid heat dissipation

Overheating is most common in video mode, especially at high frame rates.

Problem: Glitches or Frozen Screen

Symptoms: Buttons not responding, lag, black screen.

Quick Fixes:

- Turn off camera, remove battery and card, wait 10 seconds, then reboot
- Try a different memory card
- Update firmware (Menu > Firmware update)
- If persistent, reset settings (see below)

Resetting the Camera to Factory Settings

If nothing else works or you want a fresh start:

♻ How to Reset:

1. Go to **Menu > Wrench Tab > Clear Settings**

Choose:

- Clear All Camera Settings
- Clear Custom Functions (C.Fn)

You'll lose custom setups, but it's great for clearing bugs or passing the camera on.

Summary Cheat Table

Issue	Common Cause	Quick Fix
Blurry focus	Wrong AF mode	Switch to Servo/One-Shot or Face/Eye AF
Battery drains	High screen brightness, Wi-Fi	Use Eco Mode, turn off wireless
Overheating	Long 8K recording sessions	Shoot in 4K, keep cool, open LCD
Frozen camera	Software bug or bad card	Reboot + firmware update
General bugs	Bad settings or memory	Factory reset

Chapter 20

The Final Upgrade – Your Photography Workflow

Taking great photos is only the first step. Now it's time to streamline how you store, back up, and edit them like a pro without overwhelm.

Dual Card Slot Setup (Because Redundancy = Peace of Mind)

The R5 Mark II has **two memory card slots:**

- **CFexpress Type B** (Slot 1 – fast, ideal for video/RAW)
- **SD UHS-II** (Slot 2 – more affordable, still great)

Recommended Dual Slot Settings:

- **Backup Mode:** Record the same file to both cards (RAW + RAW or RAW + JPEG)

- **Split Mode:** RAW to CFexpress, JPEG to SD (for quick previews and social sharing)

- **Relay Mode:** Automatically switch to second card when the first is full

Use Backup mode for critical shoots like weddings or client work.

Organizing Files (So You Don't Go Crazy Later)

Whether you're shooting 100 photos or 10,000, good file hygiene makes editing and storage easier.

📁 Folder Strategy:

- 📅 Use date-based folders: 2025-05-10_Wildlife_Park

- 📷 Inside, separate by format:

  - /RAW

  - /JPEG

- /Exported

Tips:

- Rename files with a pattern: R5M2_2025_05_10_001.CR3

- Back up regularly to **two places**: External drive + cloud (e.g., Dropbox, Google Drive, or Lightroom cloud)

Think of your workflow like a clean kitchen, easier to cook when everything's in place.

Intro to Editing: Lightroom & Photoshop Basics

You don't need to be a Photoshop wizard. Just a few tweaks can transform your shots.

Lightroom (Best for Bulk Editing):

- Crop & Straighten: Rule of thirds grid for easy composition fixes

- Exposure & Contrast: Adjust brightness and pop

- White Balance: Fix color tone issues

- Sharpening & Noise Reduction: Clean up low-light images

- Presets: Speed up edits with one-click looks

Start with Lightroom's "Auto" button, then fine-tune.

Photoshop (Best for Fine-Tuning):

- **Remove distractions** with Healing Brush or Clone Stamp

- **Layers** let you edit parts of an image independently

- Great for **compositing, text overlays, or commercial work**

Pro Workflow in a Nutshell

1. **Shoot** (Use RAW + JPEG)

2. **Back up** to hard drive + cloud

3. **Sort** into folders

4. **Cull** (delete bad shots)

5. **Edit** in Lightroom (then Photoshop if needed)

6. **Export** for print, web, or client

7. **Archive** the final and RAW versions

Acknowledgement

To every aspiring photographer and filmmaker who dares to pick up a camera and tell a story, this book is for you. Special thanks to my family and friends for their encouragement, and to the creative community whose passion inspires me daily. Your support made this guide possible.